Polar Animals

Written by Jen Green

Illustrated by Michael Posen

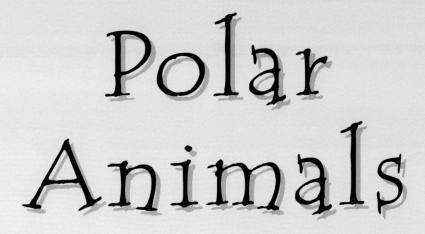

This is a Parragon Publishing Book
First published in 2000

Parragon
Queen Street House
4 Queen Street
Bath BA1 1HE, UK

ISBN 0-75254-320-2

Printed in Singapore

Produced by
Monkey Puzzle Media Ltd
Gissing's Farm
Fressingfield
Suffolk IP21 5SH
UK

Designer: Sarah Crouch
Cover design: Victoria Webb
Editor: Linda Sonntag
Artwork commissioning: Roger Goddard-Coote
Project manager: Alex Edmonds

Contents

Arctic fox

The Arctic fox keeps warm with the help of its thick fur coat and a layer of body fat.

How can animals survive the harsh polar weather?

THE POLAR REGIONS ARE THE COLDEST PLACES ON EARTH, YET SOME creatures live there all year round. Snow and ice cover the land and sea for most, or all of the year. Summers are brief and cool, with long hours of daylight. Winters are long, dark, and bitterly cold. In the Arctic, mammals such as the wolf and Arctic fox have a layer of body fat, and a thick coat of fur, to keep them warm. The fur traps air next to the animal's skin, which helps to prevent body heat from escaping.

How do Arctic foxes keep their noses warm at night?

The Arctic fox has a bushy tail up to 16 in (40 cm) long – over half its body length. When the fox is sleeping, it curls its tail around its body to cover its head and nose. The tail acts as a muff to keep the animal snug in biting winds.

Which polar animal has the longest hair?

Musk ox are large beasts related to sheep and goats. Their thick coats contain two different kinds of hair. The shaggy outer layer includes hairs up to 3 feet (1 m) long. The dense, short, woolly fur beneath gives extra warmth.

Which animal has ears adapted to keep it warm?

In hot places, such as deserts, hares have very long ears, which give off body heat to keep the animal cool. In the far north, Arctic hares have much shorter ears, which release less heat, while still giving excellent hearing.

How do birds keep warm in the polar regions?

Birds that live in the Arctic and Antarctic all year round have a dense coat of feathers. Waterproof outer feathers protect the bird against the cold and wet. Underneath, soft, fluffy down feathers help retain body heat.

How do seals keep warm in icy water?

Seals and walruses are mammals that spend most of their lives in cold water. They have a thick layer of fat, called blubber, below their skin. This fatty layer keeps them warm and well-insulated in the water. Whales and polar bears also have blubber. Whale blubber can be up to 20 in (50 cm) thick.

Harp seal

How do polar animals avoid frozen feet?

An animal's feet can be one of the coldest parts of its body, especially when, as in the polar regions, the feet touch the ice or frozen soil. Mammals, such as polar bears and Arctic foxes, have hair between their toes to prevent their paws from freezing. Some polar birds, such as the ptarmigan, a type of grouse, have feathery feet for the same reason.

Why don't polar animals need snow shoes?

Because their feet act like snow shoes! The feet of Arctic animals, such as polar bears and caribou, are broad compared to those of bears and deer from warmer regions. The broad base of the foot helps to spread the animal's weight over a wider area, so it does not sink into the snow.

The harp seal's blubber keeps it warm in cold water.

How do polar animals perform a disappearing act?

Animals all over the world have colors and

patterns on their bodies, that blend in with their surroundings. This natural disguise, or camouflage, helps them to hide from enemies and predators. In the Arctic and Antarctic, many creatures have white coats that blend in with the snow in winter. Some creatures have a different-colored coat in summer, so they can also hide when the snow has melted. In winter, the ptarmigan has mainly snow-white feathers. In summer, its mottled brown plumage helps it hide among the bare earth and rocks.

What arctic animal has two fur coats?
The stoat is a type of weasel whose camouflage varies with the seasons. In summer its brown fur blends in with the rocky landscape. In the fall it sheds its summer fur, and grows a new, thick coat of pure-white hairs. In its winter coat, the stoat is called an ermine.

Why do caribou follow in each other's footsteps?
Caribou travel as much as 1,240 miles (2,000 km) each year, on migration. To save energy, each deer treads in the footsteps of the animal in front, so it does not have to break a trail through deep snow.

Caribou herd

What color is a polar bear's skin?
Believe it or not, the polar bear's skin is black! The hollow hairs of its thick fur, reflect the light and make its skin look white.

How do baby seals make themselves invisible?
Baby harp seals spend the first weeks of their lives out of water, on the ice. Their parents have dark, mottled skins that blend in well with the water, but the young have pure white fur, which helps to hide them from predators, such as polar bears.

How do lemmings avoid catching cold?
Lemmings are furry rodents that make their homes in burrows. They pass the winter months out of the biting winds, in a network of tunnels, dug in the snow.

Long lines of migrating caribou stretch for 185 miles (300 km) or more, as the animals make their way through the barren wastes, and cross mighty rivers.

How do caribou escape the winter cold?

Some hardy animals can survive in the high Arctic all year round. Others, including caribou, visit only for the mild summer season. Caribou herds spend the summer months feeding on small plants in the tundra – the treeless lands on the edge of the Arctic Ocean. In the fall, they make their way south, to shelter for the winter in forests, called the taiga. The caribou make the same round trip each year, following the same routes. Such regular animal journeys, are called migrations.

Snowy owls

Snowy owls feed their chicks on lemmings and other small mammals.

Where does the gray whale take a winter vacation?

Many sea creatures, such as seals and whales, travel great distances when they migrate. Gray whales pass the winter months in the warm seas off Mexico. In the spring, they swim up to 12,000 miles (20,000 km) to Alaskan waters, to feed on the plentiful food there.

How does the snowy owl protect its babies?

SNOWY OWL CHICKS HATCH OUT IN A NEST ON THE GROUND IN springtime. The young and adult owls have different camouflage. The mother owl has white feathers with dark speckles, so she blends into the snowy landscape. Her chicks have mottled, gray feathers that merge with the rocks and plants near the nest.

What is the most famous Arctic animal?

RUDOLF, THE RED-NOSED REINDEER (OR CARIBOU) WHO PULLED Santa's sleigh, is certainly the best-loved Arctic animal! In Siberia and Scandinavia, tame caribou are often used to pull sleighs, and heavy loads, and people also ride on them. Throughout the Arctic, humans depend on caribou for meat, and use their skins for clothing. The name caribou probably comes from a Micmac Indian word, meaning "animal that paws through snow for food."

How big do polar bears grow?
Male polar bears can grow up to 10 feet (3 m) long, and weigh as much as 10 adult humans. The females are much smaller.

What do caribou like to eat?
Caribou is a type of deer that feeds mainly on lichen growing on rocks. This is a food that can be found all year round. In summer, the deer feed on a wide variety of plants that grow, and flourish in the long hours of daylight.

Do polar bears really live at the North Pole?
Polar bears live throughout the high Arctic, but are seldom seen near the North Pole itself. These bears catch all their food from the sea. At the North Pole, the Arctic Ocean is covered by a thick crust of ice that never melts, so the bears cannot get to the water, to swim and fish.

Which Arctic animal makes the scariest noise?
Wolves are found throughout the Arctic, living in packs. Most wolf packs contain between eight and 20 animals. They howl to communicate with other members of their group, and to warn rival wolves away from their territory.

When do grizzlies go north?
The polar bear is the only bear to spend all year on the Arctic pack ice. However, grizzly bears live in the forests that edge the tundra. In summer, they visit the Arctic to feed on berries, fish and mammals. They pass the winter hibernating in dens.

Are there cats in the Arctic?
The lynx is a large wild cat that spends most of the year in the taiga forests. It sometimes ventures into the Arctic tundra to hunt hares, voles, and lemmings.

The polar bear's scientific name, *Ursus maritimus*, means "sea bear." These mammals are often classed as sea creatures.

Wolf

Polar bear

Which animal is a champion swimmer?
Polar bears are expert swimmers and divers. They move through the water by paddling with their front paws. They have been spotted swimming as far as 200 miles (320 km) from land.

The eerie noise that wolves make has been used in many horror films!

What Arctic animal was thought to commit suicide?

Lemmings were once believed to commit suicide by throwing themselves off cliffs. Now scientists realize that these rodents breed very fast, so their numbers can rise steeply. When the area where they live becomes overcrowded and food gets scarce, large numbers of lemmings set off in search of a new food source. Many of them drown when they fall into rivers, or lakes, that lie in their path.

Lynx

The lynx's camouflaged coat allows it to creep up on victims.

What do polar bears eat?

The polar bear's favourite food is ringed seals. It hunts its victims at holes in the ice, where the seals come up for air. The bear creeps silently up to the hole, then lies in wait – for hours, if necessary. When a seal pops up to breathe, the bear kills it with a single slash of its powerful paw, and a bite to the neck.

How does the lynx go hunting?

Like other cats, the lynx has excellent sight and hearing. It hunts by silently stalking a victim such as an Arctic hare. Then it pounces on its prey with one big leap.

Why should a lemming never panic?

Arctic foxes eat a varied diet, including berries, birds, eggs, and small mammals, such as lemmings. When hunting lemmings, the fox leaps high in the air, then crashes down on top of the lemmings' burrow. The panicked lemmings run from the shelter of their home, straight into the fox's jaws.

Which animal keeps food in a "refrigerator"?

Arctic foxes sometimes cache (hide) stores of meat beneath the snow, or under rocks in summer. In cold conditions, the food stays as fresh as it would in the icebox of your refrigerator.

Which animals form a magic circle round their babies?

When musk ox are menaced by wolves, or other predators, they group themselves into a circle with their young calves safe in the middle. The adult ox face outward and fend off their attackers with their long, curved horns, until the wolves give up.

Which animal has paws like sledgehammers?

THE POLAR BEAR'S MAIN WEAPONS ARE ITS paws, which can act like sledgehammers. It can kill a human with one swipe of its massive paw. The bear's sharp canine teeth are also sometimes used to kill. Its strongest senses are smell and hearing. A hunting bear can sense prey up to 20 miles (32 km) away.

The wolverine sometimes steals prey killed by other Arctic hunters. It is so fierce that the other predator just gives up its kill.

What small creature likes a lemming for lunch?

The stoat's favourite food is lemmings. With its long, slim body, this fierce hunter can scamper down its victim's burrow, and then overcome its prey with its strong, sharp teeth.

What animal is the glutton of the Arctic?

Wolverines are large members of the weasel family, also known as gluttons. These ferocious creatures look like small brown bears. They hunt lemmings and even large mammals, such as caribou. When the wolverine makes a large kill, it gorges itself on meat, then hides the rest for later, earning a reputation for gluttony.

Which animals gang up on their victims?

WOLVES' MAIN PREY ARE CARIBOU, MOOSE, AND SOMETIMES MUSK ox. They hunt in packs and target young, weak, or injured animals. When the wolves spot a likely victim, such as a caribou calf with its mother, they spread out to surround their prey. The panicked calf tries to flee and is separated from its mother. Then the wolves run their victim down and kill it by biting its neck.

Wolverine

Plankton consists of tiny plants and animals floating near the surface of the sea. The plants use sunlight to make their food. The tiny animals feed on the plants or on each other.

Plankton

What minibeasts are food for all in the polar seas?

A WIDE VARIETY OF MICROSCOPIC PLANTS AND ANIMALS LIVE in seawater. Together, these tiny living things are known as plankton. In polar waters, plankton forms a vital food source for fish and many other sea creatures. Larger animals, such as seals, whales, and penguins, feed on the plankton-eaters, so, either directly or indirectly, all life in the polar seas depends on plankton.

Where are the hardiest minibeasts in the world?
Antarctica is a vast frozen continent where the weather is even colder and windier than in the Arctic. Conditions are so harsh, that no large land animals live there. Yet some tiny insects and spiders live and breed on the shores of Antarctica. They are some of the hardiest living things on earth.

Which is the most dreaded Arctic animal?
The mosquito is probably the most pesky polar creature. This insect feeds on animal blood. In spring, huge swarms of troublesome mosquitoes hatch out and plague the caribou herds by biting them and sucking their blood. They bite any warm-blooded animals they can find, including humans.

What minibeast can suffocate a caribou?
The warble fly lays its eggs in the caribou's skin. When maggots hatch out, they burrow into the skin and feed on the caribou's flesh. Sometimes so many maggots hatch out in a caribou's throat, that it dies from suffocation.

Why does the sea turn red and glow at night?
Krill are small shrimp-like creatures that feed on plankton. Their bodies contain light-producing organs that can give the sea a greenish glow at night. In summer, they may occur in such large numbers that they turn the water red.

How do minibeasts survive at the Poles?

Surprising as it seems, many kinds of insects and other small creatures, thrive in the Arctic. Minibeasts are cold-blooded, which means that their body temperature is only about the same as their surroundings. In winter, it is too cold for insects to be active, but they survive in the ice, or frozen soil, as eggs or larvae (young). In spring, when the ice melts, they hatch out in huge numbers, to feed on the plants that flourish in the warmer weather.

When does the sea bloom?
The warmth, and long daylight hours of the polar spring, cause plankton to multiply in vast numbers. This phenomenon is known as plankton bloom. In turn, sea creatures, that feed on plankton, also breed and flourish. So summer becomes a time of plenty in polar waters.

Krill are among the most abundant creatures on earth. All the krill in the sea would outweigh all the people on earth.

Krill

The Arctic clouded yellow butterfly's blood contains a special fluid, that prevents it from freezing.

Arctic clouded yellow butterfly

Do butterflies live in the Arctic?
We think of butterflies as delicate creatures. Surprisingly, some species thrive in the Arctic. Some Arctic caterpillars, and adult butterflies, have hairy bodies which keep them warm in freezing temperatures. Others are dark in color, because dark colors absorb the heat from sunlight quickly.

Marble plunderfish

The marble plunderfish uses the fleshy barbel on its chin like a fisherman's bait, to catch smaller fish.

When does a sea snail dry out?

Some Antarctic sea snails survive the cold by dehydrating (drying out) so the water in their body cells cannot freeze.

Why do fish need antifreeze?

Fish that live in polar seas have special features that help them survive in icy water. The arctic cod has blood that contains chemicals that work like the antifreeze in a car radiator, to prevent the fish from freezing.

Which Antarctic fish goes fishing?

The marble plunder fish has a fleshy whisker on its chin, called a barbel. To small fish, this slender tentacle looks like a wriggling worm. But when the little fish moves in, to eat the "worm," the plunder fish snaps it up.

Why are the polar seas a popular place to live?

IN THE POLAR REGIONS, FAR MORE CREATURES LIVE in the oceans than on land. In winter, the sea surface freezes over, but under the ice, the water temperature is warmer, than the air on the land. Antarctic waters contain an even greater variety of life than Arctic waters, because the ocean currents there cause nourishing minerals to well up from deep waters, to feed sea creatures.

Icefish

The icefish has pale, nearly see-through skin.

How does the icefish get its name?

Icefish can survive in very cold water, where most other fish would freeze. Unlike all other vertebrates (animals with backbones) the icefish's blood contains no red blood cells; instead, it holds a natural antifreeze. Red blood cells are normally vital because they carry oxygen round an animal's body, but the icefish can absorb oxygen direct from seawater.

Why do polar sea creatures prefer the deep?

MOST AQUATIC ANIMALS IN POLAR SEAS LIVE IN

deep water. In shallow waters, creatures that live on the seabed risk damage from floating sea-ice that scrapes along the bottom. So starfish, sponges, corals, and sea-urchins carpet the rocky bottom of the deep water. When the ice melts in summer, some of these creatures migrate inshore.

Was the Antarctic once a warm sea?

Fossil remains of ancient warm-water dwellers, such as ammonites (shelled creatures) show that the Antarctic was much warmer in prehistoric times. Ammonites were related to squid. They have long since become extinct.

When does a cucumber make a good mother?

In southern oceans, some sea creatures produce larger eggs, and take more care of their young, than similar species in warmer seas. Female sea cucumbers and sea-urchins have a special pouch, in which their young develop. This makes these young creatures much more likely to survive.

Can woodlice live underwater?

Glyptonotus antarcticus is a shelled sea creature related to woodlice. In Antarctic waters, this slow-growing crustacean can eventually become giant size, reaching 8 in (20 cm) long.

Which sharks like to chill out in the Arctic?

Most sharks live in warm or temperate waters. Greenland sharks visit Arctic seas, although they usually live in warmer, deeper waters further south. These sharks hunt squid, seals, and small whales.

Greenland sharks can grow up to 20 feet (6 m) long.

Greenland shark

Blue whale

Do killer whales deserve their name?

KILLER WHALES ARE NOT KNOWN TO KILL PEOPLE, BUT

they are deadly hunters of seals and fish. Like narwhals, they are members of the family of toothed whales. Their sharp, cone-shaped teeth are powerful weapons.

Which is the biggest polar animal?

Blue whales are the largest animals to visit the Arctic. They are the biggest whales – in fact, the largest creatures on earth. Mighty blue whales can grow up to 100 feet (32 m) long, and weigh 150 tonnes. They visit Arctic and Antarctic waters, during spring and summer, to feed. In the fall, they migrate to tropical seas, where their young are born.

Which whale uses a battering ram?

Bowhead whales have domed heads and jaws that arch upwards like a bow. The whale uses its domed head like a battering ram, to smash holes in thick pack ice, so that it can breathe. It is the only whale to swim in Arctic waters all year round.

When does a whale not mind being beached?

In southern seas, killer whales sometimes swim right up on to the beach, to snatch young seal pups. The whale cannot survive on land, so it quickly wriggles back into the water before eating its catch.

What whale sings like a bird?

Belugas are small white whales that live in the Arctic. In olden times, sailors called them "sea canaries" because of the many different sounds they make, including clicks, squeaks, bell-like clangs, and whistles.

Why was the right whale an unlucky whale?

In the days of whaling, the men of the whaling ships gave right whales their name because they were the "right" whales to hunt down and kill. Thousands of these whales were slaughtered during the 18th and 19th centuries. The whales' blubber was melted down to make oil to light lamps, and their bones were used to make umbrellas, brushes, and women's corsets.

Do killer whales have friends?

Yes – other killer whales! Killer whales live and hunt in groups called pods. Members of the pod work as a team when hunting. They spread out to surround small ice floes on which seals are resting. Then the killers charge and tip the floe from different sides. The helpless seals slide off into the water, and the hunters snap them up.

Blue whales have small eyes in proportion to their giant size, but each eye is still as large as a basketball.

Do giant whales eat mighty meals?

Which male whale likes to impress the females?

Narwhals are small Arctic whales. Males have a long, spiralling tusk growing out from their heads. Scientists are not sure what the tusks are for, but many believe their main purpose is to impress female narwhals!

BLUE WHALES DO NOT REACH THEIR GIANT SIZE BY eating large sea creatures. Instead, their meals are made up of thousands of tiny creatures, mainly krill. In place of teeth, these whales have long, thin plates of bone (baleen) hanging down from their upper jaws. These plates work like a giant sieve to strain krill, and small fish, from mouthfuls of seawater.

The narwhal's tusk is a very long tooth, growing out from the upper jaw.

Narwhal

Are seals vegetarian?

No! All seals are meat-eaters. They feed on squid, fish, shellfish, seabirds, and sometimes even other seals.

Which seal has a beard?

The bearded seal has a magnificent set of long, whiskery hairs on its nose. It uses its sensitive "beard" to feel for shellfish lurking on the seabed, and fish hiding in cloudy water.

What spotted seal snacks on penguins?

The leopard seal has a spotted skin like the leopard. Like its namesake, the seal is a speedy and deadly hunter, with large, powerful jaws. Its favourite prey are penguins and young seal pups.

Do elephant seals have trunks?

Male elephant seals have a wrinkled bag of skin on top of their nose. The bag can be blown up balloon-style, to form a kind of trunk. When inflated, the seal's trunk acts as a loudspeaker, amplifying his roars as he calls to frighten off rival males.

Which seal needs a new name?

Crabeater seals of the Antarctic do not eat crabs. They would be better renamed krill-eaters, as these small, shrimp-like animals are their main food. To catch krill, the seal speeds along with its mouth open, sieving krill from the water with the help of its jagged teeth.

Leopard seal

Leopard seals are fast movers, like the big cats after which they are named.

How do leopard seals hunt penguins?

THE LEOPARD SEAL LURKS BENEATH AN ICE FLOE ON WHICH THE BIRDS are standing. When the penguins dive into the sea, the seal darts forward and seizes a bird in its teeth. Having caught its prey, the leopard is in no hurry to feed. It beats its kill against the water, to turn the penguin inside-out, and remove its skin. It may take over an hour to finish its meal.

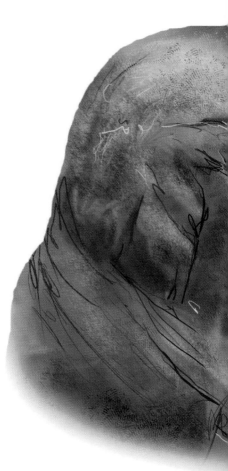

When is a seal an elephant?

The elephant seal is the heavyweight champion of the seal world. Large males may grow up to 21 feet (6 m) long and weigh up to 8,800 lb (4 tonnes). Females are much smaller. Southern elephant seals live in Antarctic waters.

Do walruses like to party?

Yes! Walruses are very sociable animals. They often gather in large, noisy groups on land, basking in the sun, or huddling together, to conserve body heat in cold weather. The walrus pack provides safety in numbers because it is more difficult for enemies to pick out weak, or young animals, from a big group.

Why do walruses have tusks?

BOTH MALE AND FEMALE WALRUSES HAVE SHARP, CURVING TUSKS up to 3 feet (1 m) long. These strong, elongated upper teeth have many uses. Walruses eat clams and other shellfish, and use their tusks to rake along the seabed to find food. The tusks are also used as levers when walruses haul their massive bodies up on to the ice. Long, curving tusks may add sex appeal when walruses are courting. The males also use theirs to gore their rivals when fighting contests to win female partners.

Walrus

The walrus' name comes from a Swedish word meaning "whale-horse."

19

Loons are expert swimmers and divers.

Loon

How does the ptarmigan defend its chicks?
Ptarmigan are ground-nesting birds, whose chicks are raised by both parents. If the young are threatened by a predator, such as an arctic fox, the parent bird flattens itself against the ground, then suddenly flies up, heading straight for its enemy's head. The chicks scatter to safety in the confusion.

What Arctic bird stores its food in a pouch?
Little auks are small, hardy seabirds that live in the Arctic all year round. They feed on plankton floating in seawater, and store their catch in a throat pouch.

What bird is known for its clumsiness?
Loons are diving birds that spend most of their lives in water. Their legs are set far back on their bodies, which makes them graceful in the water, but awkward on land. In fact, the name loon probably comes from the Icelandic word "lomr," which means clumsy.

What is the favourite food of snowy owls?
Snowy owls feed mainly on lemmings. These birds time their breeding cycle, so their chicks are born when there are plenty of lemmings to feed them. In years when lemmings are particularly abundant, the owls breed quickly and raise more chicks. When lemmings are scarce, they may not breed at all.

What Arctic bird makes friends with the polar bear?
The ivory gull is a pure-white bird that spends all year in the Arctic. Through the long polar winter, it survives by following the polar bear, as it wanders the pack ice. The gull feeds on scraps of food left by the bear and on its droppings.

What do Arctic birds like to eat?

ARCTIC BIRDS EAT MANY DIFFERENT FOODS. SEABIRDS, SUCH AS loons and puffins, dive into the seas for fish and shellfish. Wading birds and land birds, such as snowbuntings, feed on insects. Dabbling ducks eat water plants in marshy pools, and birds of prey hunt rodents. In summer, all these living things are plentiful in the Arctic, so there is enough food for these birds to rear their chicks.

How does the eider duck keep its chicks warm?
Eider ducks nest in clumps of grass on Arctic islands. The female bird plucks down feathers from her breast to line her nest and warm her chicks. Her soft, fluffy feathers are used to fill pillows.

What Arctic bird stands out in the snow?

Which bird is a deadly Arctic predator?
The golden eagle visits the Arctic in summer. This fierce predator hunts on the wing, soaring silently over the tundra in search of lemmings and ground squirrels. When it spots a victim moving in the grass below, it swoops down and seizes its unlucky prey in its powerful claws.

Most hunting birds, and many of their victims, have pale or camouflaged feathers, so they can hide in the snowy landscape. Ravens are black, so they stand out against the snow, but they are so fierce that few enemies challenge them. These clever birds live off their wits, feeding on dead meat and scraps in winter. In summer they team up in pairs to steal other birds' eggs and young. One raven distracts the parent bird, while the other steals from the nest.

Golden eagle

The golden eagle uses its sharp claws, called talons, to catch its prey – a lemming.

Why do birds fly to the ends of the Earth?

Few birds live in the Arctic or Antarctic all year round.

Most visit in spring and summer only, when the long days bring warmth and plenty. As the ice melts, flowers bloom and the air is filled with buzzing insects. Lemmings, and other rodents, emerge from their burrows, so there is plenty of food for birds to eat. Many birds also come to the polar regions, to raise their chicks in safe nesting sites, where there are few enemies to harm them. In the fall when the chicks are grown, they leave for warmer regions that do not freeze over in winter.

Where are the best polar nesting places?
Seacliffs and Arctic marshes are favourite nesting spots for migratory birds. In winter, these sites are bleak and deserted. In summer, cliff ledges and marsh rims are packed with thousands of noisy, squabbling birds.

How do birds know when to migrate?
The secrets of bird migration are still a mystery. But scientists believe that the shorter daylight hours of early fall may signal to the birds that it is time to leave.

Puffins line their burrows with tufts of grass.

Puffin

What bird lives in a rabbit burrow?
Puffins build their nests in burrows on seacliffs. The bird digs out a tunnel with its beak and feet, or takes over an abandoned rabbit burrow.

How do snow geese save energy?

Snow geese migrate great distances, flying in long lines or V-shaped groups. These formations save energy because each bird flies in the slipstream of the one in front.

Snow geese raise their young in the Arctic, then fly 2,000 miles (3,200 km) to the Gulf of Mexico for the winter.

Snow geese

How do migrating birds find their way?

Year after year, birds use the same routes as they migrate to and from the polar regions. Some young birds learn the way by flying with their parents, and following the experienced birds in front. Other birds find their way alone, purely by instinct. Birds navigate using the sun, moon and stars, and landmarks such as coastlines to establish their position. Some birds can also sense the Earth's magnetic field.

Which is the smallest Antarctic seabird?

Wilson's storm petrel is the smallest Antarctic seabird. Yet this little bird migrates almost as far as the Arctic tern, flying all the way from Antarctica, to northern Canada, and back again each year.

Why do guillemots lay pointed eggs?

Guillemots are seabirds that nest in crowded colonies on cliffs. Their eggs are pointed so if they get knocked, they roll round in a circle, and don't fall off the cliff ledge.

What bird is the long-distance flying champion?

The Arctic tern holds the record for long-distance migration. This bird nests, and rears its young in the Arctic in June, when it is light all day and all night too. Then it flies halfway round the world to take advantage of the Antarctic summer.

This map shows the amazing round trip of 25,000 miles (40,000 km) completed by the Arctic tern each year.

Which giant bird is the wanderer of the air?

THE WANDERING ALBATROSS HAS THE LARGEST

wingspan of any bird. Its outstretched wings measure 11.5 feet (3.5 m) across. The albatross uses its strong wings to soar in the winds that blow across the southern oceans. It spends almost all its life in the air, and touches down on land only to breed.

Wandering albatross

What bird brings luck and stormy weather?

In olden days, albatrosses were believed to bring rough weather. Yet sailors also thought these birds were lucky, and believed that killing one brought bad luck. *The Rime of the Ancient Mariner* is a long poem about a sailor who brings disaster on his ship, by killing an albatross. His shipmates hang the dead bird around his neck.

Which seabird lays the heaviest egg?

Albatrosses raise only one chick every two years. Their eggs are the heaviest of any seabird, weighing up to 21 oz (585 gm) and taking up to 80 days to hatch. The parent birds take it in turns to incubate the egg while it develops. The father bird spends weeks sitting on the nest.

Six species of albatross nest in Antarctica. The wandering albatross is the giant of the family.

Skua

Skuas harass other birds to
make them give up their kill.

What bird lives like a pirate?
The skua is the pirate of Antarctic
waters. It chases after other birds,
and forces them to drop or
regurgitate (bring up) their food.
These birds also gang up in teams,
to raid other birds' nests for eggs
and chicks.

How do Antarctic birds go fishing?

A NTARCTIC BIRDS USE DIFFERENT
METHODS TO CATCH FISH
in the southern oceans. Penguins and cormorants dive down
and swim underwater to overtake their prey. Terns and
shearwaters hunt from the air and plunge into the sea to
make their kill. Storm petrels flutter along the surface,
trailing their legs as if they were walking on water.
Albatrosses swim in the waves, watching out for fish in the
sea below.

**What bird is nicknamed
"stinker"?**
Giant petrels are large seabirds
related to albatrosses. They are
nicknamed "stinkers" because they
smell so awful. These birds live as
scavengers, feeding on dead seals
and other sea creatures, and
hunting fish and other birds.

**What bird will eat absolutely
anything?**
The sheathbill is a white bird, with
a horny sheath on its beak to
protect its nostrils. It survives the
freezing Antarctic winter by
grabbing whatever food it can. It
feeds on krill and dead fish, and
steals other birds' eggs and chicks
in spring. When times are hard, the
sheathbill even resorts to eating
seal and penguin droppings!

What do polar birds use to build their nests?
Polar birds build their nests with whatever materials they can lay their
beaks on! On the barren shores of Antarctica this includes seaweed,
feathers, lichen, moss, mud, straw, and even stones.

What do shags use to glue their nests together?
Blue-eyed shags nest in colonies on the coasts of Antarctica. The weather is
so harsh there that these birds may use their nests all year round, not just
to raise their chicks. Shags build large nests of seaweed, moss, and lichen,
stuck together with the only glue available, their own droppings!

How do emperors keep their eggs warm?

Where do penguins live?

Penguins live only in the Southern Hemisphere. There are 16 different species. Most live in the far south, in Antarctica, or on islands in the south Pacific or Atlantic Oceans. A few species occur further north, and one even lives on the remote Galapagos Islands, near the Equator.

E MPEROR PENGUINS DON'T BUILD nests, and an egg left on the ice would freeze in seconds. When the female lays her egg, the male scoops it up onto his feet, and covers it with a warm flap of skin. He incubates the egg for two months during the worst winter weather, until the female returns to help feed the newly-hatched chick.

Which penguin is the deepest diver?

The emperor penguin can dive to depths of 870 feet (260 m), and stay below the surface for 18 minutes. The emperor is the largest penguin, growing up to 4 feet (1.2 m) tall.

Why is a penguin like a cork?

To leave the water, penguins swim up to the shore at speed, and then pop out of the water like a cork from a bottle, to land on the ice feet-first. They sometimes use this technique to escape from predators, such as leopard seals.

Emperor penguin

Out on the windy pack ice, male emperor penguins incubate their eggs.

When is a penguin like a toboggan?

On land, penguins walk with an ungainly waddle. When descending snowy slopes, they toboggan along on their stomachs, pushing with their feet for extra speed.

How did chinstrap and Adelie penguins get their names?

The chinstrap penguin has a narrow line of black feathers running under its white throat, like a chinstrap. Adelie penguins were named by the French explorer Dumont d'Urville, who discovered Antarctica in 1840, after his wife Adelie.

Why do penguins steal pebbles?

Penguins breed in large, noisy colonies called rookeries. Some make no nests, but chinstrap and Adelie penguins build nests of pebbles. Aggressive chinstraps sometimes steal one another's pebbles, or take over Adelie nests.

Which penguin lays its eggs on ice?

Most Antarctic penguins breed on land in summer. But emperor penguins breed in winter, out on the freezing ice. In April, they trek 60 miles (100 km) to traditional nesting sites, where they mate. Then the females head back to the sea, and the males are left to incubate the eggs.

Can penguins fly?

PENGUINS ARE FLIGHTLESS BIRDS, BUT THEY ARE STRONG SWIMMERS AND divers. Underwater, they "fly" along, using their wings as flippers to push against the water. When swimming at speed, they often leap out of the water and dive back in again, an action known as porpoising.

How do penguins keep warm?

In the harsh Antarctic climate, penguins keep warm with the help of a layer of fatty blubber, and two layers of feathers. Their outer coat of tiny oily feathers overlap to keep the down feathers below, dry. Penguins also have partly feathered beaks and small feet, so only a small part of their body is exposed to the freezing air and ice.

Adelie penguins

Seven species of penguin live in Antarctica. These are Adelies.

One guillemot threatens another by spreading its wings and pointing skywards with its beak. Its opponent backs down by preening its feathers or bowing with outstretched wings.

How do guillemots keep the peace?

GUILLEMOTS REAR THEIR CHICKS ON CROWDED CLIFF LEDGES. EACH breeding pair has only a tiny space to feed its chick. In these overcrowded conditions, fights between rival birds would quickly lead to chicks or eggs falling to their deaths. Squabbling guillemots resolve their differences by using ritual movements that show threat and submission, so physical fights are rare.

How does a male caribou prove its strength?
Fall is the mating season for caribou. The males strut and roar, then charge at one another and lock antlers. The deer that wins this test of strength also wins the herd of females.

What makes polar bears fight?
Polar bears live mostly solitary lives, roaming the lonely pack ice. Females breed only once every three years. If two male bears come across a female who is ready to mate, a fight breaks out. The males wrestle fiercely in the snow, for only the victor will get the chance to mate.

What bird dances to attract a mate?
Sandhill cranes migrate to the Arctic to breed. The male and female birds court one another by performing "dances" with many different steps. The birds dip down and bow low, make skipping movements and leap up to 20 feet (6 m) in the air.

Which penguin flirts with its eyebrows?
Rockhopper penguins have tufts of feathers above their eyes that form spectacular "eyebrows." Males and females court by squawking at one another and waggling their heads to show off their impressive tufts.

When is wrinkled skin a turn-on?
Like elephant seals, male hooded seals have a bag of loose, wrinkled skin on top of their heads. During the breeding season, the male attracts the females, and warns rivals away, by roaring and inflating his "hood" into a big black balloon.

Guillemots

What is a beachmaster?

Southern elephant seals breed on islands off Antarctica. Rival males fight fiercely, rearing up and tearing at each other's necks with their teeth. The seals' necks are padded with thick skin and fat, but the animals are soon bloody. The victor becomes "beachmaster," winning a stretch of beach and his own group, or "harem," of females.

Elephant seals

What makes a wolf leader of the pack?

WOLF PACKS HAVE A STRICT SYSTEM, IN WHICH THE STRONGEST animals dominate the weaker ones. Usually, only the strongest male and female wolf mate. Whenever they meet, wolves show their place in the hierarchy through their body language. Top wolves stand proud with pricked ears, bared teeth and tails held high. Junior wolves cower with their ears flat, and their tails between their legs.

Which animal can walk a great distance at only two days old?

Caribou calves are born in June in the Arctic. In minutes, they are able to stand on their long legs and wobble around. In a day or two, they are ready to keep up with the herd as it moves on in search of fresh grazing. In the fall, the calves travel south with the herd to the sheltered forests.

Where do narwhals go to raise their young?

Narwhals spend most of their lives swimming near the edge of the Arctic pack ice. But they migrate to the fiords of Greenland and Scandinavia to give birth. The young spend their first weeks in the warmer waters of these sheltered inlets, where they are fairly safe from enemies such as killer whales.

Why do Brant geese raise their young in a hurry?

BRANT GEESE ARRIVE TO BREED IN ARCTIC MARSHES AS SOON as the ice melts. The young must be raised before the brief summer ends, so the birds lay their eggs at once. The eggs hatch out quickly, and the chicks grow fast. In less than two months the young birds are ready to take off with their parents as they head back to warmer climes.

Which animal drinks milk underwater?

Like other mammals, young whales feed on their mother's milk. The female's nipples are located on the underside of her body, so the calf must take a deep breath, and dive down below its mother to drink.

What animal has several dens?

Ringed seal pups are born in an ice den, scraped out by their mother with her flippers. The female makes several dens, so she can move her pup to safety if it is threatened.

What seabird takes a long time to raise its chick?

On the windswept shores of Antarctica, wandering albatrosses take ten and a half months to rear their young. The parent birds take turns to feed their chick on an oily, foul-smelling mixture made of regurgitated fish.

Young Brant geese learn to swim soon after they hatch out.

Polar bear cubs spend two years suckling their mother's milk and learning hunting skills. Then they are on their own.

Polar bear cubs

What games do polar bear cubs play?

Young polar bears are curious and very playful. They chase one another, slide down snowy slopes, and tussle in the snow. Through play, they learn the skills they will need to survive, once they are weaned.

Which chicks are looked after in a creche?

Emperor penguin chicks hatch out in July – midwinter in Antarctica. They spend about eight weeks riding round on their parents' feet, clear of the ice. Then, they are left in "creches" with other young penguins, while their parents go off to hunt for food.

Where are polar bear cubs born?

POLAR BEARS GIVE BIRTH IN MIDWINTER TO UP TO three cubs. The young are born in a snug snow den, hollowed out by their mother. They live in the den for three months, feeding on her rich milk. Then, the female breaks through the snow plugging the entrance to the den, and the cubs scamper out into the icy world for the first time.

Brant geese

31

Index